A Day in the Life Of

An Ant

Ruby Tuesday Books

Ruth Owen

Published in 2025 by Ruby Tuesday Books Ltd.

Copyright © 2025 Ruby Tuesday Books Ltd.

All rights reserved. No part of this publication may be reproduced in whole or in part, stored in any retrieval system, or transmitted in any form or by any means, electronic, mechanical, photocopying, recording, or otherwise, without written permission from the publisher.

Editor: Mark J. Sachner
Design: Tammy West
Production: John Lingham

Photo Credits:
Alamy: 11 (Biosphoto), 12 (Hakan Soderholm), 19 (Antje Schulte/Ant Life); iStock: 7 (Istvan Balogh), 8–9 (samuiboy), 13B (Lucia Ghetti); Nature Picture Library: 10 (Rod Williams), 15 (Stephen Dalton); Science Photo Library: 17 (Clouds Hill Imaging); Shutterstock: Cover (Dmitry Dolhikh), 4 (Rupendra Singh Rawat), 5 (Ksenia Lada), 6 (Yeti Studio & PaulrommerSL), 13T (Inga Nielsen), 16 (ABO Photography), 20 (Photo Fun), 21 (Maximillian cabinet), 22 (Dwi Yulianto, Anton Kozyrev, & Irina Kozorog), 23 (Catcher of Light Inc, irin-k, & Jirasak Kaewtongsorn), 24 (K.IvanS); Kamil Stajniak: 14; Superstock: 18 (Matt Cole).

ISBN 978-1-78856-440-3

Printed in Poland by L&C Printing Group

www.rubytuesdaybooks.com

CONTENTS

Meet a Busy Worker Ant 4

Glossary 22

Index 24

Meet a Busy Worker Ant

It's morning in a garden.

A black garden ant and her family leave their underground nest.

She is a worker ant, and her job is to find food.

Antennae

Worker ant

The ant uses her two **antennae** to smell for food.

The ant finds a sugary sweet that was dropped by a human.

She scurries back to the nest.

Her body makes a special smelly **trail**.

Other worker ants follow her trail to eat the sweet.

Ants like to eat sugary food.

The little worker ant finds a dead millipede.

Worker ants

This **HUGE** meal needs teamwork!

The ant makes a smelly trail back to the nest.

Dead millipede

A team of ants comes to carry the millipede back to their home.

Thousands of ants live in the nest.

A dead damselfly

The worker ants must find lots of food for their family.

When they need a drink, the ants sip water from raindrops.

The ant makes a trail from something sweet on the garden's patio.

JAM!

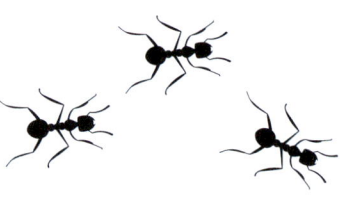

The ant has two **stomachs**. One is for the food she eats.

The stomachs are in here.

A rotting apple

Her second stomach is for food she carries back to the nest.

Back at their underground nest, the ants sick up the food they collect.

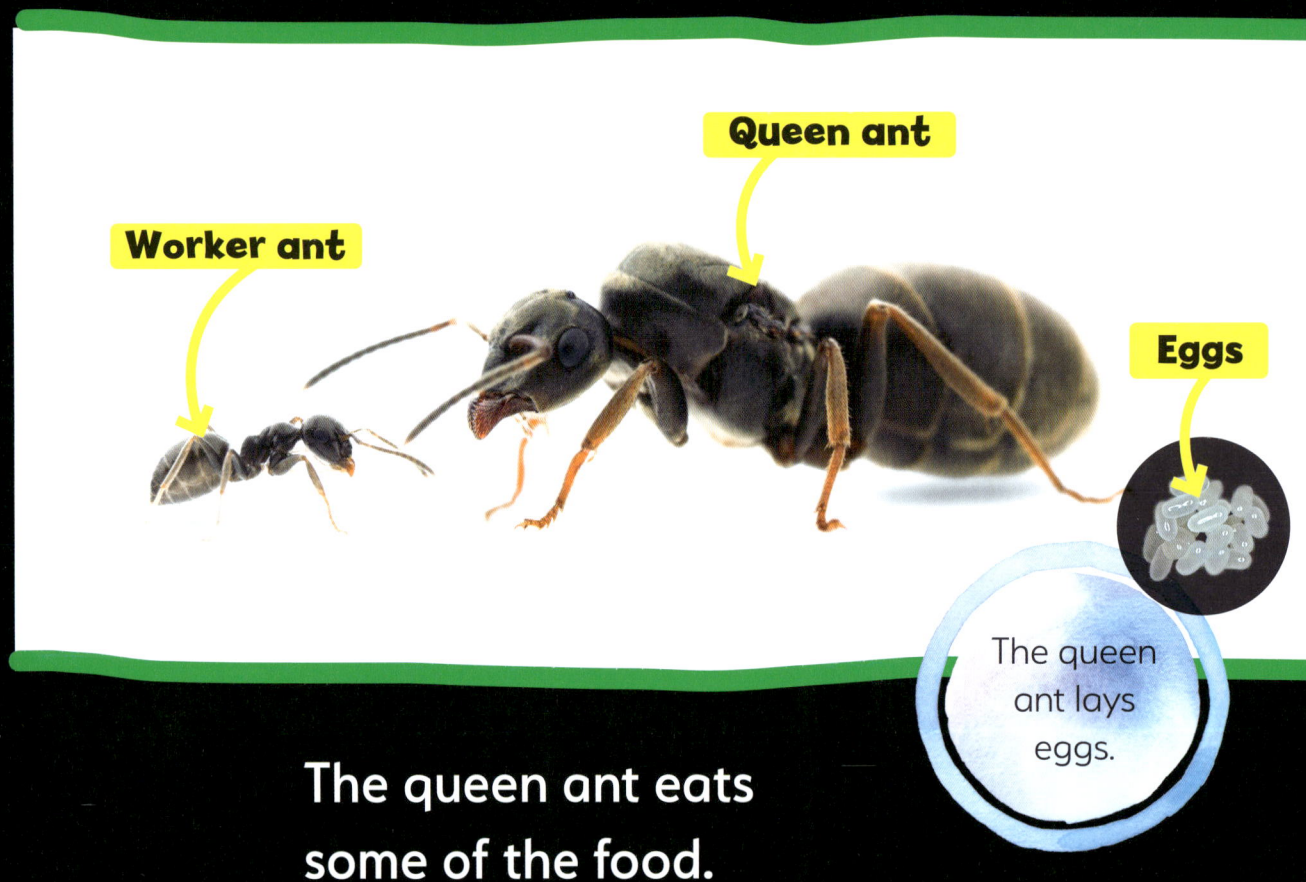

Worker ant

Queen ant

Eggs

The queen ant lays eggs.

The queen ant eats some of the food.

She is the mother of all the ants.

Workers in the nest also feed the food to the baby ants.

Worker ant

Larva

A baby ant is called a **larva**.

The busy worker ant has another way to get food.

With the other workers, she takes care of little **insects** called aphids.

Aphids suck a sugary juice called sap from plants.

Mouthpart

A close-up picture of an aphid

The worker ant tickles an aphid with her antennae.

This makes the aphid poo sweet stuff called honeydew.

The worker ant gobbles up the honeydew.

Ladybirds and other insects eat aphids.

The worker ants protect their aphids from these **predators**.

The ants bite the hungry ladybird.

When evening comes, the little ant goes underground.

Ant nest

Tomorrow will be another busy day!

Glossary

antennae
Two long body parts on the head of an insect. Antennae may be used for touching, smelling and tasting.

insect
A tiny animal with six legs. Ants, ladybirds, beetles and bees are all insects.

larva
A young insect that hatches from an egg.

Larva

predator
An animal that hunts and eats other animals.

stomach
A bag-like body part where food goes after it's eaten by a person or animal.

An ant's stomachs are in here.

trail
A pathway that can be followed. A trail might be marked with signs. Some animals use smell to mark a trail.

Index

A
antennae 5, 18
aphids 16–17, 18–19, 20

E
eggs 14

F
food 5, 6–7, 8–9, 10, 12–13, 14–15, 16, 19

H
honeydew 18–19

L
ladybirds 20
larvae 15

N
nests 4, 6, 9, 10, 13, 14–15, 21

Q
queen ants 14

S
stomachs 13

T
trails 6–7, 9, 12